LIONS

Amy-Jane Beer

Grolier
an imprint of
SCHOLASTIC
www.scholastic.com/librarypublishing

ST. CHARLES SCHOOL LIBRARY

Published 2009 by Grolier
An imprint of Scholastic Library Publishing
Old Sherman Turnpike, Danbury,
Connecticut 06816

© 2009 Grolier

All rights reserved. Except for use in
a review, no part of this book may be
reproduced, stored in a retrieval system,
or transmitted in any form, or by any
means, electronic or mechanical, including
photocopying, recording, or otherwise,
without prior permission of Grolier.

For The Brown Reference Group plc
Project Editor: Jolyon Goddard
Picture Researcher: Clare Newman
Designers: Dave Allen, Jeni Child, Lynne Ross,
    Sarah Williams
Managing Editors: Bridget Giles, Tim Harris

Volume ISBN-13: 978-0-7172-8028-5
Volume ISBN-10: 0-7172-8028-4

Library of Congress
Cataloging-in-Publication Data

Nature's children. Set 4.
    p. cm.
    Includes bibliographical references and
index.
    ISBN 13: 978-0-7172-8083-4
    ISBN 10: 0-7172-8083-7 ((set 4) : alk. paper)
    1. Animals--Encyclopedias, Juvenile. I.
Grolier (Firm)
    QL49.N385 2009
    590.3--dc22
                         2007046315

Printed and bound in China

PICTURE CREDITS

Front Cover: **Shutterstock**: Jonathan Heger.

Back Cover: **FLPA**: Frans Lanting; **Nature
PL**: Peter Oxford; **Shutterstock**: Chris
Fourie, Snowleopard.

Nature PL: Terry Andrewartha 33, Laurent
Geslin 46, Vincent Munier 37, Anup Shah 18,
38; **NHPA**: E. Hanumantha Rao 10;
Shutterstock: EcoPrint 9, Jeff Goldman 29,
Adam Harner 45, Jonathan Heger 17, Emin
Kuliyev 4, 26–27, Chad Littlejohn 21, Jasenka
Luksa 6, Louie Schoeman 2–3, 13, Kristian
Sekulic 41, Shutterspeed Images 5,
Snowleopard 22, Johan Swanepoel 30,
Stefanie Van Der Vinden 42, Adam Ward 14;
Still Pictures: BIOS/Michel Denis-Huot and
Christine Pictures 34.

Contents

Fact File: Lions 4

Extended Family 7

Ancient Lands 8

Out of Africa 11

Paws and Claws 12

Mane Attraction 15

Sharp Senses 16

Proud Pride 19

Roar Power 20

Lion Talk 23

Touchy-feely 24

Lazy Days 25

Feature Photo 26–27

Team Effort 28

Meaty Feast 31

A Free Lunch? 32

Sticking Together 35

Out with the Old 36

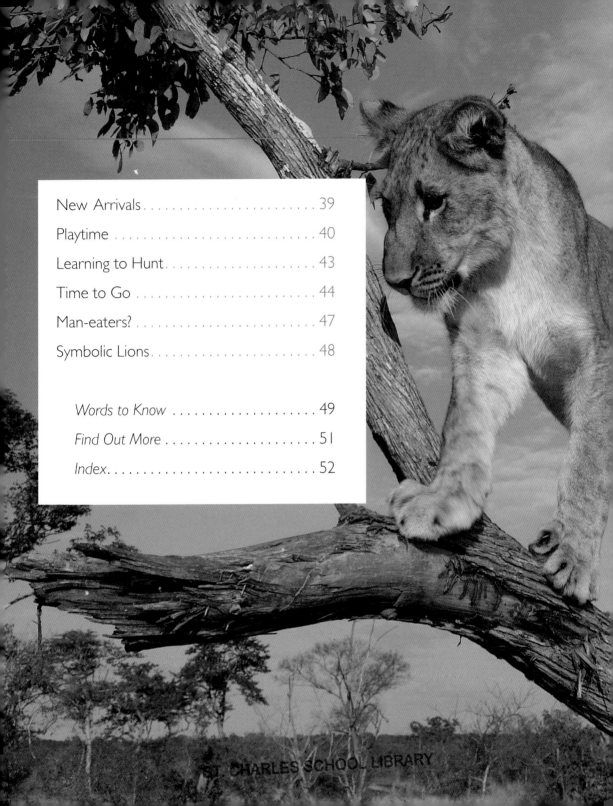

New Arrivals 39
Playtime 40
Learning to Hunt 43
Time to Go 44
Man-eaters? 47
Symbolic Lions 48

  Words to Know 49
  Find Out More 51
  Index 52

FACT FILE: Lions

Class	Mammals (Mammalia)
Order	Carnivores (Carnivora)
Family	Cats (Felidae)
Genus	Lions, tigers, leopards, and jaguars (*Panthera*)
Species	Lion (*Panthera leo*)
World distribution	Africa, south of the Sahara Desert; some live in the Gir Forest in India
Habitat	Savanna grasslands and light woodlands
Distinctive physical characteristics	Large gold to tawny cat; male has mane of longer hair around neck and shoulders; cubs are lightly spotted
Habits	Live in family groups; active day and night, but spend a large part of the time resting; females hunt in teams
Diet	Meat, mostly that from hoofed animals such as antelope and zebra

Introduction

Often called the "King of the Beasts," there is something about the lion that commands respect. Perhaps that is because these mighty **predators** do not seem to fear anything, and that they work together in a way that humans admire. These powerful cats fight over the same sorts of things that humans fight over, too, such as land, relationships, and the right to protect and provide for their families. The shaggy hair, or **mane**, around the head and neck of adult males also gives these big cats a kingly appearance.

Male lions start to grow a mane at about two years old.

Cats have an extra sense organ in their nose, called a Jacobsen's organ, that lets them "taste" smells.

Extended Family

Lions are big cats—very big cats! In some ways they are similar to pet cats. They have a strong, supple body. They have sharp hooked claws that they can draw up, or retract, into their feet. They also have teeth perfect for catching **prey** and eating meat. And just like pet cats, they like nothing better than snoozing in the sunshine. But there are some important differences between a lion and an average house cat. Size is the most obvious difference. A lion weighs up to 550 pounds (250 kg)! A house cat rarely weighs more than 16 pounds (7 kg).

Size is not the only difference between these distant relatives. The black pupils of a lion's eyes are always round. In a small cat, they close up into slits in bright light. Small cats usually crouch to eat their food, while lions lie down. Only lions have shaggy hair, or a mane, around their head. They can roar, too, while small cats can only wail, mew, hiss, or purr.

Ancient Lands

Lions once lived in North and South America. They did not live in zoos. They lived in the wild and hunted wildlife. These ancient lions went extinct after the last ice age—about 10,000 years ago—when the landscape changed from open plains to dense forests. Scientists who study **fossils** sometimes discover their bones preserved in caves in North and South America.

Lions also used to live in Europe and Asia. They died out gradually as people spread out and began hunting many of the same prey. Eventually, there was little food left in the wild for the lions to eat. The last lions in Europe died out in Roman times—about 2,000 years ago—but a few survived in North Africa and the Middle East until the start of the 19th century.

Lions live for 10 to 15 years in the wild. In captivity, they can reach 25 years old.

The Indian, or Asiatic lion, once ranged across India, Turkey, and Persia (Iran).

Out of Africa

Most people know that lions live in Africa. What they might not know is that lions live in India, too. Unlike, the plains-dwelling lions of Africa, Indian lions live in forests. Male Indian lions don't grow a big mane like that of African male lions. They do have a mane, but it is short. That way, the mane doesn't get snagged in the forest undergrowth.

Indian lions are very rare. About 100 years ago there were only about 100 of them left. Their chances of survival have now improved since their home—a place called the Gir Forest—was made into a special reserve. There are now about 300 Gir lions in the wild and another 150 or so Indian lions in zoos around the world.

Paws and Claws

Lions have big feet called paws. Each paw has four toes. The front paws have a fifth mini toe called a **dewclaw**. A lion has a large sharp claw on each of its other four toes. The claw is made of **keratin**, which is the same material as human fingernails. However, a lion's claw is much thicker and stronger, and it is shaped into a very sharp hook. Like most other cats, the lion can retract, or draw in, its claws inside the soft pad of its toes.

When retracted, the claws don't catch on the ground when the lion walks normally, so they don't get worn or damaged. Retracted claws also allow lions to be very gentle when they choose. But don't be fooled—those deadly weapons can spring out in an instant to slash at an enemy or hook into the body of prey.

Young lions sometimes climb trees to escape from flies on the ground or to enjoy a cool breeze.

A lion's mane darkens or gets richer in color as the animal ages.

Mane Attraction

Adult male lions have a mane—a thick shaggy cape of long hair that grows from around the face, down the neck and throat to the shoulders and chest. The mane looks magnificent, but it also serves some important purposes. Firstly, a shaggy mane makes a male lion look much bigger than it actually is. That way the lion is less likely to be attacked. Sooner or later, however, most male lions have to fight—usually other male lions—and when they do, the mane helps protect their throat from the claws and teeth of their enemies.

Sharp Senses

Lions have excellent senses. Their eyesight and hearing are particularly fine-tuned. They can see just as well in the dark as they can by day. Their large furry ears twitch back and forth in order to pick up even the smallest of sounds.

Lions have a good sense of smell, too. They use scent to recognize one another. They leave scent marks around their **territory** so other lions know they are around. Finally, lions have a fine sense of touch. The long whiskers on their face act as feelers when the lions are moving about in pitch darkness. With all these senses sharply aware of the world around them, it's not often that a lion is taken by surprise.

Lions, like other predators, have forward-facing eyes. That helps them judge distances accurately when attempting to pounce on prey.

A pride of lions might range up to 200 square miles (500 sq km) in search of prey.

Proud Pride

Lions are the only cats that live in an extended family group. This group is called a **pride**. There can be up to 6 males, 12 females, or **lionesses**, and their young, or **cubs**, in a pride. Typically, most groups are smaller than that, however.

The lions in a pride work together and look out for one another. The lionesses do most of the hunting, and they care for and teach the cubs. The big adult males appear to do little more than lie around, not helping much with anything else. However, the males have a vital job to do. While they might seem to be simply lying around, they are in fact keeping a close watch out for predators. They are the guardians of the pride, and they will fight, sometimes to the death, to protect the lionesses, cubs, and their territory. For this reason, male lions are very large—up to twice the size of the females.

Roar Power

Only big cats like lions, tigers, and leopards can roar—and the lion is better at roaring than all the others. A lion's roar is a deafening sound. It can be heard even by human ears more than five miles (8 km) away. Often the members of a pride roar together. The sound drifts over the landscape so other lions nearby know there is a strong pride in the area. Roaring together seems to help the pride bond—similar to sports fans that support the same team by cheering together. Roaring is hard work. Only lions more than one year old can manage it. Most roaring sessions happen at night.

Lions usually roar after sundown. Both adult males and lionesses can roar.

A lion cub calls for its mother.

Lion Talk

In addition to roaring, lions make a lot of other, much quieter sounds. They growl, moan, hiss, snarl, huff, and puff. Cubs often squeak and mew like small cats. All these sounds are used to communicate with other lions, and usually their meaning is clear. A hiss, growl, or snarl is a warning. A sharp "woof" is a sound a lion makes when it is startled. Quiet moans and grunts show contentment. One sound lions can't manage so well is purring. They can make a sort of purr in short bursts, but it's not the same as the continuous purr one might hear from a contented house cat.

Touchy-feely

Lions in a pride are often very affectionate with one another. They spend a lot of time lying side by side, nuzzling and rubbing their heads and faces together and licking one another. All these actions help the bond between the pride members stay strong. Adults often greet cubs by giving one of their cheeks a rub or a lick.

Before every hunt there is usually a sort of team pep talk, where the lionesses rub faces and make grunting and puffing sounds as though encouraging one another. If the hunt doesn't go well, it can take a while for the scattered pride to regroup, but when they do, there is another round of face rubbing and moaning sounds that say, "Welcome back. Better luck next time!"

Lazy Days

There is little a lion likes more than a long nap. As long as there is plenty of food to go around, lions will spend up to 21 hours each day sleeping. That leaves just three hours each day for other things, such as hunting, eating, drinking, and **nursing** cubs.

However, lions are rarely in such a deep sleep that they become completely unaware of their surroundings. Their eyes might be shut, but their ears and other senses are still alert. Whenever possible, they sleep in the shade to avoid overheating, and usually they sleep close together. That way if any of them notice something wrong and stirs, the others wake up, too.

A male lion can grow to 10⅕ feet (2.5 m) long and stand 4 feet (1.2 m) tall at the shoulder.

Team Effort

The lionesses in a pride work as a team to hunt. Usually one lioness tries to sneak up on the prey, creeping along on her belly so she is completely hidden by the tall grass. At the same time the others fan out around the prey so whichever way it runs one of the lions will be there to block its path. Lions can run at 36 miles (58 km) per hour in short bursts.

The lionesses are very skilled, but their prey are so alert and fast that most teams only manage to make a kill in about one out of every four hunts. They usually kill the prey by grabbing its neck in their jaws and suffocating it. But lions have to be careful. A well-aimed kick from a zebra or giraffe can shatter a lion's jaw or skull, killing the hunter instantly or dooming it to starvation. When successful, the lions share the kill with the rest of the pride. The biggest share usually goes to the males and the scraps to the cubs.

Three lionesses plan their attack on a herd of buffalo.

A lion's teeth are only suitable for grabbing, stabbing, and slicing—not chewing.

Meaty Feast

Lions eat almost nothing but meat. Their digestive system cannot cope with any other type of food. If a lion tried to eat leaves or plants it would get very little energy from its meal and probably become ill. Meat is much easier to break down, or digest, than plant matter, but it is also much harder to find. After all, grass and leaves don't run away!

While most plant-eating animals spend part of every day feeding, lions might only get to feed once every few days. They don't usually have safe places to store their food, so they have to eat as much as they can immediately. For that reason lions have teeth that allow them to slice up meat quickly, and a large stomach that can stretch to hold very big meals. A large lion needs to eat an average of 15 pounds (7 kg) of meat each day.

A Free Lunch?

It was once thought that lions had a terrible time being pestered by hyenas. It is common to see lions trying to eat, while a pack of hyenas hangs around, making a racket and constantly trying to make off with some of the lions' kill. In fact the truth is often the other way around. Some lions find it much easier to let the hyenas hunt, then simply steal their meat. That way, the lions don't have to chase down prey themselves or risk getting injured in the hunt. It's a smart move on the lions' part, but understandably the hyenas don't like it one bit. It's no wonder the hyenas make such a fuss while the lions feed!

Hyenas and lions are arch enemies. Hyenas are one of the main killers of lion cubs.

A male lion will fight to the death to protect his pride and territory.

Sticking Together

In most prides, all the adult lionesses are related to one another. The adult males are related, too, but only to one another—not to the females. Sometimes, there is just one adult male lion in a pride. However, that is unusual because, unless he is extremely strong, it is easy for other males outside the pride to team up and defeat him. A team of young males—usually brothers or cousins—is called a **coalition**. The team roams the land until it finds a pride it can take over and make its own. Usually, that means fighting with older males and driving them away. These fights can be terrible, and some lions die from their injuries.

Out with the Old

When new males take over a pride, it is a time of great upheaval for all the other members. Older cubs have to run for their life and never come back. Often the lionesses try to leave, too. The youngest cubs are doomed—the new males in charge kill them. It seems very cruel, but the new males know they might only be able to keep the pride for a few years. If they are to have families of their own, they need to start immediately. Amazingly, the lionesses soon seem to forget their loss. Before long there are new cubs and the males become gentle, doting fathers and uncles. Life in the pride settles down once again.

The bond between lionesses in the same pride is very strong.

A cub knows to keep quiet and not wriggle when its mother moves it.

New Arrivals

Lions can breed at any time of the year. It takes three to four months for a **litter** of cubs to develop inside their mother. There can be anything from one to six cubs in a litter, but usually there are three or four. Often, several litters are born at about the same time in a pride.

The lionesses share the work of looking after them. They will happily nurse cubs that are not their own. All the adults in the pride help protect the young cubs from enemies. Lions from other prides and hyenas will both kill cubs if they are given the chance. Young cubs have a spotted coat that provides a bit of **camouflage**. Their first reaction when frightened is to crouch down in the grass and hide.

Playtime

Young cubs are never still for long. They are full of life and spend a lot of time playing. They chase one another and play-fight. They try to get the adult lions to join in, too. To begin with, they are very nervous around the big males. But eventually their high spirits seem to melt the heart of even the grumpiest old lion, and they might be allowed to climb all over him. Sometimes, they need to be reminded of their place. However, even if a cub gets cuffed or snapped at, the warning is usually gentle.

The cubs of a pride huddle together for safety while the lionesses hunt.

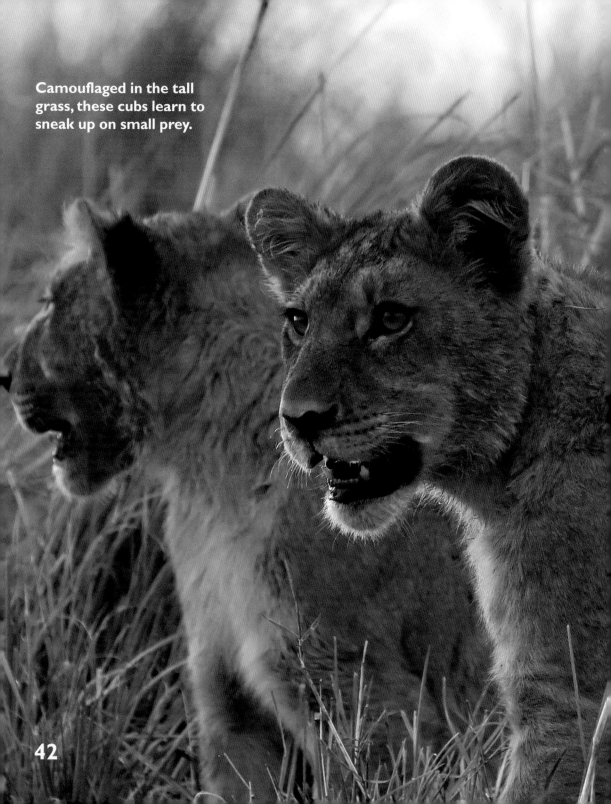
Camouflaged in the tall grass, these cubs learn to sneak up on small prey.

Learning to Hunt

Many of the games cubs play are hunting games. They practice pouncing on anything from insects and clumps of grass to other cubs and even the tufted tail of the big male lions in the pride.

When they are about six months old, the cubs are given pieces of meat or small prey to practice their stalking and killing moves on. At about ten months the cubs join in on real hunts. To begin with, they are far too excitable, and many hunts are ruined by their impatience. The cubs attack too soon, giving the prey far too much warning. That must be very frustrating for the adult lionesses! At the age of about a year and a half, the young lions are able to play a useful part in most hunts.

Time to Go

By the time they are two or three years old, young male lions become a bit of a nuisance in the pride. The adult males begin to get impatient with them. There are many squabbles and sometimes some serious fights.

Soon the young males will be old enough to breed. To do that, the young males must move away and find a pride in which the lionesses are not related to them. Usually, the young males leave the pride with their brothers and cousins of the same age. They spend a few years roaming the land, hunting and growing bigger and stronger. They will then start looking for a pride to take over.

If a young male lion that has left the pride is on his own, he might try to form a coalition with unrelated young males.

A pride of lions makes use of the road in Kruger National Park, South Africa.

Man-eaters?

Traditionally, lions have had a bad reputation as dangerous animals that kill livestock and sometimes even eat people. But, these days, such attacks are rare. When people and livestock are killed by lions, it is usually because the lions have nothing else to hunt. That happened a lot in the past when people built farms in the lions' territory and drove out all the lions' natural prey. Many thousands of lions were killed to protect people and their animals.

Lions were also killed for sport. Thankfully, now in most places, there is more respect for wild lions, although a certain amount of illegal hunting, or poaching, still occurs. In fact, in game reserves and national parks, the lions are welcome. Tourists that come to see lions and other wild animals on safari bring money into the area.

Symbolic Lions

People have always admired lions. There are thousands of stories about them from many different cultures. In Roman legend, Androcles (AN-DRUH-KLEEZ) helps a lion by removing a thorn from its paw. The stories by C. S. Lewis feature a country called Narnia ruled by a great lion called Aslan.

Lions also appear in art as statues and in many famous paintings. They are used as symbols of strength and nobility by everyone from European kings and queens to sports teams and advertisers. Hundreds of brands use a lion in their name or emblem. Let's hope that our admiration for this mighty cat helps the **species** survive well into the future.

Words to Know

Camouflage   Patterns and colors that make an animal difficult to see against a particular background.

Coalition   A group of male lions—usually related—that work together to take over and look after a pride.

Cubs   Baby lions.

Dewclaw   A small toe on the back of the foot of some animals that is not used for walking.

Fossils   The remains or traces of ancient animals or plants that have been preserved in the ground or in rocks.

Keratin   A tough substance that makes up claws, fingernails, and hooves.

Lionesses   Female lions.

Litter   A group of baby animals born together.

Mane   Shaggy hair around the head, neck, shoulders, and on the chest of a lion.

Nursing   Feeding young mammals on milk from teats.

Predators   Animals that hunt other animals.

Prey   An animal that is hunted by other animals.

Pride   A family group of lions.

Species   The scientific word for animals of the same kind that can breed together.

Territory   An area that an animal or group of animals defends as its own private space.

Find Out More

Books

DK Publishing. *Cat.* Eyewitness Books. New York: DK Publishing, 2004.

Markle, S. *Lions.* Animal Predators. Minneapolis, Minnesota: First Avenue Editions, 2004.

Web sites

Creature Feature: Lions

www.nationalgeographic.com/kids/creature_feature/0109/lions.html

Fun facts, a video, an audio clip, and more.

Lions

www.enchantedlearning.com/subjects/mammals/lion/coloring.shtml

Facts and a printout of lions to color in.

Index

A, B, C
Africa 8, 11, 46
breeding 39, 44
camouflage 39, 42
claws 7, 12, 15
climbing 13
coalition 35, 45
communication 22, 23
cubs 19, 22, 24, 25, 28, 36, 38, 39, 40, 41, 42, 43

D, E, F
dewclaws 12
ears 16, 25
enemies 12, 15, 39
extinction 8
eyes 7, 17, 25
eyesight 16
feeding 25, 31
feet 7, 12
fighting 5, 15, 19, 34, 35, 44
fossils 8

H, I, J
habitats 11
hearing 16
height 27
hunting 19, 24, 25, 28, 29, 32, 43, 44, 47
hyenas 32, 33, 39
Indian lion 10, 11
Jacobsen's organ 6
jaws 28

K, L, M
keratin 12
length 27
leopards 20
life span 9
lionesses 19, 21, 24, 28, 29, 35, 36, 37, 38, 39, 41, 43, 44
litter 39
males 5, 15, 19, 21, 27, 28, 34, 35, 36, 40, 43, 44, 45
mane 5, 7, 11, 14, 15

N, P, R
nursing 25, 39
paws 12
pet cats 7
play 40, 43
play-fighting 40
poaching 47
predators 5, 17, 19
prey 7, 8, 12, 17, 18, 28, 42, 43
pride 18, 19, 24, 28, 34, 35, 36, 37, 39, 41, 43, 44, 46
purring 23
roaring 7, 20, 21, 23
running 28

S, T, W
scent marks 16
size 7, 19
sleep 7, 25
smell 6, 16
stomach 31
teeth 7, 15, 30, 31
territory 17, 34
throat 15
tigers 20
touch 16
walking 12
weight 7
whiskers 16